Khaled Alsaleh
Graeme Paton

Bioaccessibility of Cu and risk assessment models

AF135582

Khaled Alsaleh
Graeme Paton

Bioaccessibility of Cu and risk assessment models

Risk assessment and bioaccessibility

LAP LAMBERT Academic Publishing

Impressum / Imprint

Bibliografische Information der Deutschen Nationalbibliothek: Die Deutsche Nationalbibliothek verzeichnet diese Publikation in der Deutschen Nationalbibliografie; detaillierte bibliografische Daten sind im Internet über http://dnb.d-nb.de abrufbar.
Alle in diesem Buch genannten Marken und Produktnamen unterliegen warenzeichen-, marken- oder patentrechtlichem Schutz bzw. sind Warenzeichen oder eingetragene Warenzeichen der jeweiligen Inhaber. Die Wiedergabe von Marken, Produktnamen, Gebrauchsnamen, Handelsnamen, Warenbezeichnungen u.s.w. in diesem Werk berechtigt auch ohne besondere Kennzeichnung nicht zu der Annahme, dass solche Namen im Sinne der Warenzeichen- und Markenschutzgesetzgebung als frei zu betrachten wären und daher von jedermann benutzt werden dürften.

Bibliographic information published by the Deutsche Nationalbibliothek: The Deutsche Nationalbibliothek lists this publication in the Deutsche Nationalbibliografie; detailed bibliographic data are available in the Internet at http://dnb.d-nb.de.
Any brand names and product names mentioned in this book are subject to trademark, brand or patent protection and are trademarks or registered trademarks of their respective holders. The use of brand names, product names, common names, trade names, product descriptions etc. even without a particular marking in this work is in no way to be construed to mean that such names may be regarded as unrestricted in respect of trademark and brand protection legislation and could thus be used by anyone.

Coverbild / Cover image: www.ingimage.com

Verlag / Publisher:
LAP LAMBERT Academic Publishing
ist ein Imprint der / is a trademark of
OmniScriptum GmbH & Co. KG
Heinrich-Böcking-Str. 6-8, 66121 Saarbrücken, Deutschland / Germany
Email: info@lap-publishing.com

Herstellung: siehe letzte Seite /
Printed at: see last page
ISBN: 978-3-659-67483-9

Zugl. / Approved by: Aberdeen, University of Aberdeen, Scotland/UK, 2008

DEDICATION

I dedicate my work to my family. A special feeling of gratitude to my wife for her patience and understanding. I dedicate this work and give special thanks to my parents, Father and Mother, for encouragement and instilling the importance of hard work and higher education.

I also dedicate this work to my friend JOE who has supported and encouraged me throughout my work, and for helping me develop my technology skills.

CONTENTS

LIST OF TABLES

LIST OF FIGURES

CHAPTER 1
Introduction

A range of risk assessment techniques have been developed to assess the relationship between soil pollution and human exposure. Human exposure pathways to soil contaminants (e.g. metals) include soil ingestion, dermal and vegetable ingestion. It has been proposed that the major exposure route of metal in soil, especially for children, is oral ingestion (Paustenbach, 2000) which occurs as a result of direct consumption either intentionally or unintentionally from unwashed vegetables or poor personal Hygiene (Intawongse and Dean 2006). Children who consume between 50–200 mg of soil daily are more sensitive than adults to metal exposure. (Calabrese et al., 1997).

In terms of a child's exposure and risk assessment to human health, most risk assessment models estimate the external exposure (intake) to contaminant (usually expressed as mg/kg bw/day) rather than the total internal exposure (uptake) to contaminant capable of causing measured harm (Versantvoort et al., 2005). Metals (e.g. copper) are sequestrated in soil due to a combination of adsorption, partitioning or chemical bonding; they are tightly bound leading to reduced bioavailability to humans. Additionally, some metals are more readily sorbed from ingested soil than from the medium characterised in the toxicological or epidemiological study (EA, 2005). Therefore, it has been suggested that both the bioaccessibility and the bioavailability of metals in soils should be taken into account for a less conservative and a more accurate risk assessment (Ljung K. et al., 2007).

In a toxicological context, bioavailability is defined as the fraction of a compound which is sorbed into systemic circulation (Ruby et al., 1999; Paustenbach 2000). *In vivo* (live animal) studies are used to produce bioavailability data based on the measurement of the contaminant in the blood or organs of the animals. Although the results of these studies have been shown to be reasonable analogues (Casteel *el al.* 1997), difficulties in the interpretation of bioavailability data have been found when applied to human health due to the difference between physiology of human and the experimental animals being used (Rudy *et al.* 1999).

Up to date, it is assumed that the bioavailability and bioaccessibility of metals from an ingested soil or food are equal to that from the matrix (liquid or animal food) implemented in toxicity studies (Ljung K. et al., 2007). This assumption might lead to overestimate of internal exposure to the contaminant explained be that the bioaccessibility is less than 100% means the internal exposure to the contaminant is lower than the external exposure (intake) (Versantvoort et al., 2005). Therefore, in vitro digestion models have been developed as a useful tool for health risk assessment by determining a relative bioaccessibility by comparing the contaminant bioaccessibility from the ingested matrix with that from the matrix used in the toxicity studies (Versantvoort et al., 2005).

In vitro digestion models focussing on oral bioaccessibility are used to estimate the potential human health risks from metals by measuring their bioaccessibility in soil or food samples. The bioaccessibility is defined as bioaccessible fraction of a compound which is soluble in the gastrointestinal tract and is thus available for absorption (Ruby et al., 1999; Paustenbach 2000). Most *in vitro* gastrointestinal models are static models which are the easier to use for simulating transit through the human digestive tract by sequential exposure of the samples to mimic mouth, gastric, and small intestinal conditions (Intawongse and Dean, 2008; Oomen *et al., 2002*). Few groups of researchers have performed dynamic gastrointestinal models (e.g. TNO gastrointestinal models, TIM, Netherlands) which mimic the gradual transit of a number of gastrointestinal solutions through the simulated physiological conditions in the digestive tract (Intawongse and Dean 2008; Oomen *et al., 2002*).

Comparison of five different in-vitro gastrointestinal models (Oomen *et al.* 2002) indicated that there are variations in bioaccessibility values among them interpreting by that each model has been developed for a particular purpose with specific contaminants and matrix and used a specific procedure. The only two methods which have been validated against *in vivo* models are the Physiologically Based Extraction Test (PBET) method for lead, Pb (Rudy *et al., 1996*) and *in vitro* gastrointestinal (IVG) method for arsenic, As (Rodriguez *et al.* 1999). The PBET test is a two stage sequential extraction at 37 °C using different enzymes to simulate gastric and smell intestine compartment. This test does not use food in the extraction to mimic fasting conditions. It has been validated with Sprague-Dawley rat model for lead (Pb), and rabbit and primate models for arsenic (As) in contaminated soil and media (Rudy et al., 1999). Additionally, this method is widely used for Pb bioaccessibility due to the strong correlation with the results of in vivo models, but for other contaminants (e.g. As, Cd, Cr and Hg), it was less successful in validating the available results with confidence due to the lack of in vivo and in vitro data (Intawongse and Dean, 2006). While the IVG method has been validated for As in contaminated soils and waste materials against in vivo (a immature swine) model with the use of food in the extraction procedure (Rodriguez et al. 1999). The results indicated a well correlation and good agreement for both stomach and intestinal extraction phases. Therefore, the validation of other methods is incomplete because of the lack of in-vivo data, and the lack of appropriate certified reference materials (CRMs) for bioaccessibility studies of humans (Intawongse and Dean, 2006). A number of studies have shown a variation of bioaccessibility (and oral bioavailability) values (Table 1).

Table 1 the differences in the bioaccessibility values for As, Cd and Pb

Element	Soil type	Bioaccessibility	Citation
As	contaminated soils - gossan soils	12.1-16.4%	Juhasz A.L. 2007
	railway soils	11.2- 74.7%	
	Flanders (light sandy loam)	6-95%	Oomen et al. 2002
	Oker 11 (sandy loam)	1-19%	
	Montana 2711	10-59%	
	Mining and smelting material	4% to 43%	Rodriguez et al. (1999)
	alluvial soil and mineral beneficiation waste	11.2% in stomach 18.9% in small-intestine	Williams et al. 1998
	mine waste site	11% in small intestine	Davis et al. 1992
	Roadside in Anaconda	12% in small intestine	
	Historic soil I	69.4% gastric component	Ellickson et al. 2001
	Historic soil II	76.1% gastric component	
Cd	Flanders (light sandy loam)	7-92%	Oomen et al. 2002
	Oker 11 (sandy loam)	5-92%	
	Montana 2711	6-99%	
	Ten contaminated soils	63% and 38.2% in fasting and fed condition in gastric phase	Schroeder et al., 2003
	Ten contaminated soils	39.1% and 12.9% in fasting and fed condition in intestinal phase	
Pb	Bunker Hill soil	2% to 33% (fasted condition) 7% to 29%. (fed condition)	Van de Wiele et al., 2007
	Flanders (light sandy loam)	4-91%	Oomen et al. 2002
	Oker 11 (sandy loam)	1-56%	
	Montana 2711	3-90%	
	mine waste site	5.6% in small intestine	Davis et al. 1992
	Roadside in Anaconda	0.18% in small intestine	
	Historic soil I	65.9%. in small intestine	Ellickson et al. 2001

Recently, an interest in the use of in vitro methodologies has been increased and developed to investigate the human oral bioavailability of metals from food and soil under fasted and fed condition (Ruby et al., 1999; Oomen et al., 2003; Versantvoort et al., 2005). However, the *in vitro* digestion model (Oomen et al., 2003) was used in the current study due to its simulation for the digestion conditions in three compartments, including mouth, gastric and intestinal tracts.

In-vitro methods for metals bioaccessibility in soil and food samples have been recently reviewed (Intawongse and Dean, 2006). It is important in the development of

7

such methods to consider some features which influence gastrointestinal extraction such as temperature, gastric and intestinal pH, solid-to-solution ratio, chemical compositions, food constituents, mixing and incubation time (Intawongse and Dean 2006). Physiologically, the temperature of the human body is normally 37 °C so all extractions should take place at this temperature to avoid any effects on enzyme activity and chemical characteristics like solubility. The residence time and gastric pH in the stomach compartment are based on fasting and fed conditions (8 to 15 min with gastric pH 1-2 and 0.5 to 3 h with gastric pH 2-5, respectively). The typical chemical compositions of saliva, gastric, duodenal and bile juices used in these models as well as enzymes can be adapted from Wittsiepe *et al.* (2001) or from Rotard *et al.* (1995). However, under fed condition, the compositions of these juices should be changed due to the changes in digestive secretions in human gastrointestinal tract when eating food (Versantvoort et al., 2005).

To increase relevance, several methods include the addition of food constituents to augment the *in vitro* approach (Versantvoort et al., 2004, 2005). The presence of food might decrease the uptake of contaminant (such as lead) (Ruby *et al.* 1996); however, other studies showed that the inclusion of food can increase the bioaccessibility of some metals (Rodriguez et al. 1999; Oomen *et al.* 2002). Many studies have shown a variation of metal bioaccessibility values in soil and food samples (Juhasz A.L. 2007; Van de Wiele *et al.*, 2007; Oomen *et al.* 2002; Ellickson et al. 2001), indicating that the use of these methods requires an understanding of the nature and characteristics of metals and soil as well as their biological, chemical and physical conditions (Intawongse and Dean, 2008). It is also considered that the particle sizes of soil, <2 mm (Oomen *et at.* 2002), <250 μm (Davis *et al.* 1997) and <150 μm (Casteel *et al.* 1997) might influence the metal bioaccessibility due to the limitation of metal access to binding sites (Ljung *et al.,* 2007). To standardise bioaccessibility results, it is preferred that soils should be sieved to <250 μm particle size after oven drying at <45°C due to the consideration of an incidental ingestion and adherence of soil to children fingers (Rodriguez *et al.* 1999), as well as its routinely use in electron microprobe investigations supporting human risk assessments (Davis *et al.* 1997).

Copper is an essential nutrient element with a minimum daily requirement of 0.02 to 0.08 mg/kg-day (WHO, 1996). Therefore, a TDI for copper cannot be lower than the levels required for essentiality. TDIs provided by animal studies have been shown to be lower than the daily requirements for protecting human health. Thus, the National Institute of Public Health & Environmental Protection (RIVM, the Netherlands) has assessed the non-cancer oral toxicity data for copper and its TDI on the daily intake of

8

human which is 0.14 mg/kg-day as an upper limit (Baars AJ et al. 2001). Although the precise levels of intake or exposure to copper are still not known for both human and animal studies, high exposure doses of copper might cause serious damage to body e.g. liver and kidneys (Bremner and Beattie 1995). Additionally, Copper is representative of a wider range of heavy metals in its chemical properties and behaviour in soil and prevalent in wastes applied to land. Hence, in vitro digestion model can be used as an assessment tool and thus can estimate the intake exposure of copper to human (especially children).

The main objectives of this study were (1) to assess the bioaccessibility of copper in different soils spiked with copper using *in vitro* digestion method under fated condition; (2) to compare Cu bioaccessibility under fasted condition with that under fed condition for the historical soil; (3) and to test the hypothesis that risk assessment models for copper in different soils are more conservative than empirical data.

CHAPTER 2

Materials and methods

2.1. Sampling procedure and sample preparation:

Four soils, previously amended with copper at a range of concentrations were weathered by sequential leaching. The soils had measured concentrations of (mg kg^{-1}): Culbin (190, 320, 540, 821 and 920); Brechin (30, 175, 540, 1890 and 5120); Insch (20, 180, 505, 2100 and 5200); and St. Fergus soils (20, 520, 890, 9840 and 17000). The Aberlour soil (historical contamination) was contaminated with copper (58, 94, 58, 98 and 110 mg kg^{-1}). All soils were thoroughly mixed prior to use and air-dried in secured container for 24 hours at 25 °C. Soils then were sieved to obtain the particle size fraction of 180 μm. Physicochemical characteristics of the four laboratory spiked soils are given in Table 2 including soil pH, organic matter (OM) content, C:N Ratio, texture and iron oxide concentration which determine Cu binding to soil (Dawson et al., 2006).

2.2. Procedure of *in vitro* digestion model:

2.2.1. Preparation of Synthetic Fluid:

Organic and inorganic solutions used for preparing the synthetic fluids under fasted condition were adapted from Oomen et al., 2003, and for fed conditions were adapted from Versantvoort et al., 2004 (Table 3). The pH of saliva, gastric, duodenal and bile under fasted condition were in range of (6.5 ± 0.2), (1.07 ± 0.07), (7.8 ± 0.2) and (8.0 ± 0.2), respectively. Whereas under fed condition were (6.5 ± 0.2), (1.30 ± 0.02), (8.1 ± 0.2) and (8.2 ± 0.2), respectively. All fluids were prepared and put in water bath at a constant temperature (37°C). The chemicals used for the digestive fluids in Table 3 were sourced from Sigma (St. Louis, MO, USA), BDH Chemicals Ltd. (Poole, UK), and Fisher Scientific UK Ltd. (Loughborough, Leicestershire). An element standard of Cu was purchased from Fisher Scientific UK Ltd. (Loughborough, Leicestershire). Chemical solutions and digestive juices were prepared with deionised (DI) water (2.2 MΩ cm) from the Milli-Q^{TM} Water System.

Solutions were prepared at the beginning of experiment at room temperature and maintained at 4 °C except CaCl$_2$.2H$_2$O, NH$_4$Cl and Na$_2$HPO$_4$ which were freshly prepared on the day of use (Oomen et al., 2003). The constituents such as α- amylase, mucin, pepsin, Bovine serum albumin (BSA) and lipase were also kept at 4 °C, whereas pancreatin was stored at -4 °C. The artificial saliva, gastric juice, duodenal and bile juices were also prepared on the day of use, and then kept in water bath at 37 °C (body temperature) before digestion. Milli-Q water was used for the preparation of all these chemicals and fluids.

Table 2 The fundamental soil properties which determine Cu binding to soil, include soil pH, organic matter (OM) content, C:N Ratio, texture and iron oxide concentration (Dawson et al., 2006).

Site	pH in CaCl$_2$	%Clay	%Sand	%Silt	%OM	C:N Ratio	Cd	Cu	Zn	Ox. Al	Ox. Fe	Ox. Mn	Ox. Si
Brechin	6.04	15.00	70.18	14.82	5.50	12.77	<0.25	15.05	56.38	2843	4581	200	702
Culbin Forest	3.65	2.00	98.00	0.00	1.68	56.00	<0.25	0.39	1.92	398	305	3.0	173
Insch	5.43	12	69	19	4.32	13.21	<0.25	8.21	17.34	1457	5421	190	678
St. Fergus	6.81	5.00	93.00	2.00	8.62	11.13	<0.25	9.35	25.91	282	3397	97.2	535

(Ox.) = oxalate. Cd, Cu, Zn and oxalate (Ox.) extractable elements (Al, Fe, Mn and Si) are all elements and oxalate extractable elements have been measured in mg kg^{-1}.

Table 3 Constituents and concentrations of the synthetic juices of the *in vitro* digestion used under fasted and fed condition.

	Saliva	Gastric Juice	Duodenal Juice	Bile
Inorganic solution	10 ml KCl 89.6 g/L [c] 10 ml KSCN 20 g/L [a] 10 ml NaH$_2$PO$_4$ 88.8 g/L [a] 1.7 ml NaCl 175.3 g/L [c]	15.7 ml NaCl 175.3 g/L [c] 3.0 ml NaH$_2$PO$_4$ 88.8 g/L [a] 9.2 ml KCl 89.6 g/L [c] 18 ml CaCl$_2$.2H$_2$O 22.2 g/L [b] 10 ml NH$_4$Cl 30.6 g/L [a]	40 ml NaCl 175.3 g/L [c] 40 ml NaHCO$_3$ 84.7 g/L [b] 10 ml KH$_2$PO$_4$ 8 g/L [c] 9 ml CaCl$_2$.2H$_2$O 22.2 g/L [b] 6.3 ml KCl 89.6 g/L [c] 10 ml MgCl$_2$ 5 g/L [c] 180 µl HCl 37% g/g [c]	30 ml NaCl 175.3 g/L [c] 68.3 ml NaHCO$_3$ 84.7 g/L [b] 4.2 ml KCl 89.6 g/L [c] 10 ml CaCl$_2$.2H$_2$O 22.2 g/L [b]
Organic solution	8 ml urea 25 g/L [c]	3.4 ml urea 25 g/L [c] 10 ml glucose 65 g/L [a] 10 ml glucuronic acid 2 g/L [a] 10 ml 33 g/L glucoseamine hydrochloride [a]	4 ml urea 25 g/L [c]	10 ml urea 25 g/L [c]
some constituents	1 g BSA [c] **Fasted condition** 10 ml Na$_2$PO$_4$ 57 g/L [c], 1.8 ml NaOH 40 g/L [b], 145 mg α-amylase [a], 15 mg uric acid [a], 50 mg mucin [a] **Fed condition** 10ml NaSO$_4$ 57/l [c], 290 mg α-amylase [a], 15 mg uric acid [a], 25 mg mucin [a]	1 g BSA [c] **Fasted condition** 8.3 ml HCl 37% g/g [c], 1 g pepsin [a], 3 g mucin [a] **Fed condition** 6.5ml HCl 37%g/g [c], 2.5 g pepsin [a], 3 g mucin [a]	1 g BSA [c] **Fasted condition** 3 g pancreatin [a], 0.5 g lipase [a] **Fed condition** 9 g pancreatin [a], 1.5 g lipase [a]	1.8 g BSA [c] **Fasted condition** 200 µl HCl 37% g/g [a], 6 g bile [a] **Fed condition** 150 µl HCl 37% g/g [a], 30 g bile [a]
pH	6.5 ± 0.2 6.5 ± 0.2	1.07 ± 0.07 1.07 ± 0.07	7.8 ± 0.2 7.8 ± 0.2	8.0 ± 0.2 8.0 ± 0.2

a – Chemicals provided by Sigma (St. Louis, MO, USA).

b - Chemicals provided by BDH Chemicals Ltd. (Poole, UK).

c – Chemicals provided by Fisher Scientific UK Ltd. (Loughborough, Leicestershire).

2.2.2. Soil-to-solution ratio:

Soil-to-solution ratios (g:ml) were adapted from Oomen et al. 2003 for compartmental extraction which is that 0.6 g dry matter soil to 58.5 ml digestion juice (1:98), and the ratios of digestive juices- saliva, gastric juice, duodenal juice, and bile- were 1:1.5:3:1, respectively. Whereas, soil-to-solution ratios (g:ml) for sequential extraction have been changed to be 0.6 g soil to 46.5 ml digestion juice (1:78) with keeping up the ratios of digestive juices (1:1.5:3:1). This change was due to the use of conical centrifuge tubes (50 ml). For fed condition, 4.5 g of food was added to 0.6 g of contaminated soil, and ratios of digestive juices were 1:2:2:1 for saliva, gastric juice, duodenal juice, and bile, respectively (Versantvoort et al., 2005).

2.2.3. In vitro extraction:

Sequential and compartmental extractions were carried out using three control samples and three replicated samples which include 6 g of soil (dry weight) under fasted and fed condition. The extraction of fed state was only for Aberlour soil by adding 4.5 g of an infant formula to 0.6 g. An infant formula (mixed vegetables with noodles and chicken, HiPP UK Ltd.) was introduced as a standard meal. This infant formula was chosen because it is organic meal for children from seven months, had already been cooked and needed only to be eaten.

All digestion tubes for fasted and fed conditions were rotated over and under in a rotating chamber at about 55 rpm at 37 °C for the mouth, stomach, and small-intestine compartments. The over and under rotation method was used because it has appeared to obtain more reproducible results than a horizontal shaking method for determining phthalate mobilization from baby toys into artificial saliva (Brandon *et al.,* 2006). The tubes were centrifuged at 2,500 G for 10 min at 37 °C. At the end the chyme (the supernatant) was yielded for chemical analysis.

2.2.3.1 Sequential extraction:

Sequential extraction was carried out by adding 6 ml of artificial saliva to 0.6 g of soil. This mixture was rotated for 5 min at 37 °C and then 10.5 ml of gastric juice was added, and again the mixture is rotated for 2 h. Finally, 24 ml and 6 ml of duodenal and bile juices, respectively, were added simultaneously, and the mixture was rotated for another 2 h. The chyme was thus separated after centrifuging for 10 min (2500 G, 37 °C) and 0.5

ml of 5 M HNO_3 was added to chyme. All samples were thus kept at 4 °C until analysis. Following the same processes above, fed condition was carried out by adding 6 ml of artificial saliva to 4.5 g food and 0.6 g soil. Then 12 ml gastric, 12 ml duodenal, 6 ml bile juices and 2 ml bicarbonate solution were added.

2.2.3.2. Compartmental extraction:

Compartmental extractions were carried out for mouth, stomach and small intestinal compartments, in order to investigate the bioaccessibility of copper from their matrix during transit in the gastrointestinal tract.

- *Mouth compartment:* 9 ml of saliva was added to 0.6 g of soil for fasted condition; while 6 ml of saliva was added to 4.5 g of the infant formula and 0.6 g of soil. The mixture was rotated for 5 min at 55 rpm (37 °C). The digestion tubes were centrifuged for 10 min at 2,500 G; then the chyme was yielded and all samples were kept at 4 °C after the addition of 0.5 ml of HNO_3 for analysing.

- *Stomach compartment:* 13.5 ml of gastric juice was introduced to 0.6 g of soil for fasted condition and 12 ml of gastric was added to 4.5 g of the infant formula and 0.6 g of soil. The mixture was rotated for 2 h, and then the chyme was separated after centrifuging for 10 min at 2,500 G (37 °C) and 0.5 ml of 5 M HNO_3 was added. All samples were then kept at 4 °C prior to chemical analysis.

- *Small intestinal compartment:* 27 ml and 9 ml of duodenal and bile juices, respectively, were introduced simultaneously to 0.6 g of soil under fasted condition; while under fed condition, 12 ml of duodenal, 6 ml of bile juices and 2 ml bicarbonate solution were added simultaneously to 0.6 g of soil. The mixture was then shaken for 2 h and centrifuged for 10 min at 2,500 G (37 °C) to yield the chyme. Finally, 0.5 ml of 5 M HNO_3 was added to all digestion tubes which kept at 4 °C until analysis.

2.3. Copper analysis:

2.3.1. Determination of copper by FAAS

Determinations of copper contents in the chyme (aqueous form) obtained after centrifugation were carried out using flame atomic absorption spectrometry (FAAS) equipped with a deuterium lamp as a system of background correction for measurements. The hollow-cathode lamp was operated at 23 mA, and the wavelength was 324.8 nm. Calibration of instrument was carried out by different Cu standards which were diluted with 6.34 ml (0.1 M HNO_3).

2.3.2. Calculation of Cu bioaccessibility:

The measurements of Cu bioaccessibility are normally reported as relative bioaccessibility which is expressed as a percentage and then calculated per digestion (Oomen et al., 2002).

$$\text{Bioaccessibility (\%)} = \frac{\text{contaminant determined in chyme mobilized from soil } (mg\ kg^{-1})}{\text{contaminant present in soil before digestion } (mg\ kg^{-1})} \times 100$$

2.4. Statistical analyses:

Normality distribution was test in order to identify the suitable statistical test. Thus, non-parametric statistics for Two-way ANOVA (Friedman) was used as an alternative to determine whether there are significant differences among series of Cu bioaccessibility % in the gastrointestinal tract, and copper concentration in soil. Paired t-test (paired two sample for means) were used to investigate whether there was a significant difference between the mean of bioaccessibility in the compartmental extraction and between Cu bioaccessibilities under fasted and fed condition. Minitab 15 software was used throughout.

CHAPTER 3

Results

3.1. Copper concentration in soils:

The experiments were carried out on spiked soil samples (Culbin, Brechin, Insch and St. Fergus) and historically impacted soil (Aberlour). Table 4 summarises the measured concentration in the soils relative to the initial amendment. A prolonged period of leaching and soil ageing had taken place to the metals into the soils. This caused the greatly variable recorded concentrations of Cu in each of the soils. Furthermore in the cased of St. Fergus, much of the Cu became intimately associated with the organic carbon on the soil which was the main soil component adopted in this soil.

Table 4 Cu concentrations in Aberlour soil (Historical contamination); the amount of Cu spiked in the four soils; and Cu Concentrations after spiking with soils.

Historical contamination of Cu $(mg\ kg^{-1})$	Cu spiked in soil $(mg\ kg^{-1})$	Cu Concentrations measured in soil particle sizes (180 μm) after period of spike $(mg\ kg^{-1})$			
Aberlour		Culbin	Brechin	Insch	St. Fergus
58	0	190	30	20	20
94	50	320	175	180	520
110	250	540	540	505	890
58	1000	821	1890	2100	9840
98	2500	920	5120	5200	17000

3.2. Effect of Cu concentration on bioaccessibility in each soil:

The bioaccessibilities of Cu were investigated in each contaminated soils with different Cu concentrations using sequential and compartmental extractions (Table S1). The sequential extraction is whole system that the digestive fluids (saliva, gastric and Duodenal with Bile –D+B) were added to soil sample one after another, while compartmental extractions is the extraction for each digestive fluid separately. Based on the sequential extraction, Cu bioaccessibilities were variable with total Cu concentration. There was a significant effect of Cu concentration on the bioaccessibility for Aberlour (historical contamination) and spiked Culbin soils (Friedman Test, $P = 0.004$ and 0.009, respectively) (Table 5). Cu concentration in Culbin soil of 190 and 920 mg kg^{-1} corresponded to bioaccessibility values of 32.10 ± 0.02 and 52.62 ± 0.04 %, respectively. While in historical soil the Cu bioaccessibility was 105.69 ± 0.03 % at 58 mg kg^{-1} and 36.28 ± 0.01 % at 110 mg kg^{-1}. However, no significant effect on the bioaccessibility were found for Brechin ($P = 0.092$), Insch ($P = 0.463$) and St. Fergus soil ($P = 0.126$). In Insch soil, the Cu bioaccessibility in sequential extraction was 26.13 ± 0.01 % and 25.73 ± 0.03 % at Cu concentration in soil of 180 mg kg^{-1} and 5200 mg kg^{-1}, respectively. It was observed that the Cu bioaccessibility (%) did not always increase with increasing total Cu concentration, and in some soil the bioaccessibility became slightly decline with high Cu concentration (Fig. 1 & 2).

Table 5 P-values of Friedman test for Cu concentration on the bioaccessibility % for Aberlour (historical contamination) and spiked soils

	Bioaccessibility %				
	Aberlour soil -fasted	Culbin soil	Brechin soil	Insch soil	Fregus soil
Cu (mg kg^{-1}) blocked by Extraction	0.004	0.009	0.092	0.463	0.126
Extraction blocked by Cu	0.004	0.002	0.019	0.011	0.007

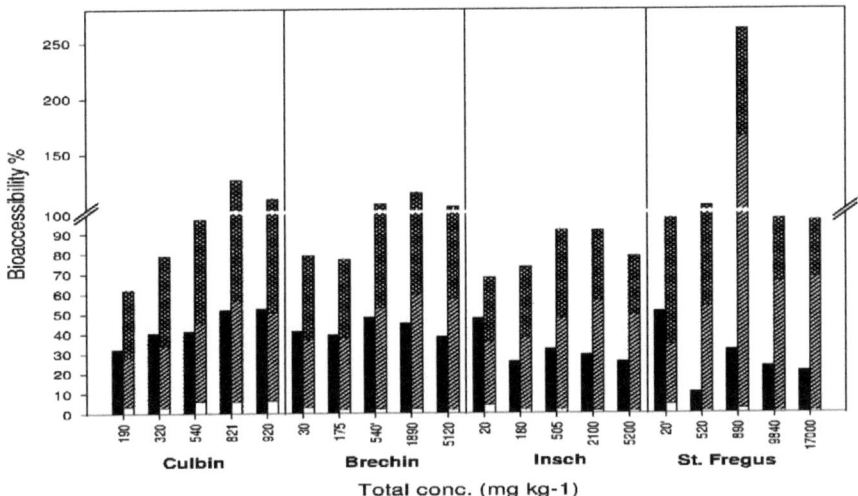

Fig.1 Cu bioaccessibilties for Culbin, Brechin, Insch and St. Fergus soils (spiked soils) in both sequential and compartmental extractions under fasted condition:

■ Sequential extraction; ☐ Saliva; ⧄ Gastric; ▩ Duodenal with Bile.

Sequential and compartmental extractions for Historical Aberlour soil contaminated with copper under fed condition

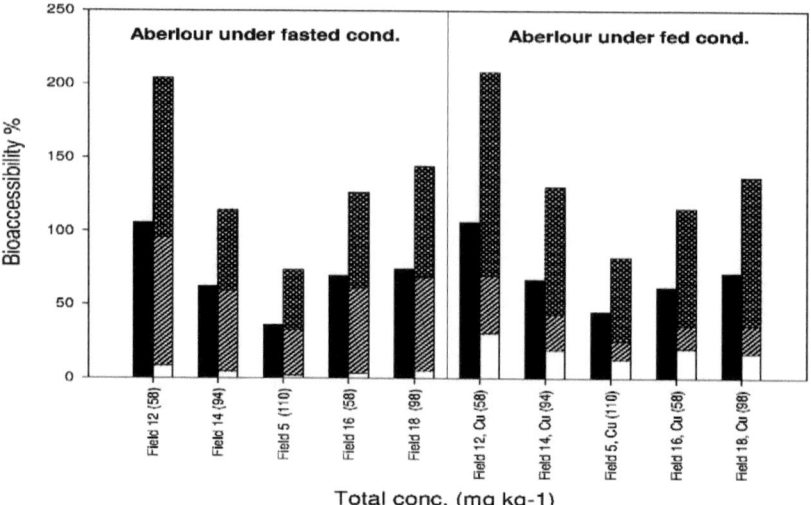

Fig.2 Cu bioaccessibilties for Aberlour soils (historical contamination) in both sequential and compartmental extraction under fasted and fed condition:

  ▬ Sequential extraction; ☐ Saliva; ▨ Gastric; ▩ Duodenal with Bile.

3.3. Compression of Cu bioaccessibility in different extraction:

The effect of both sequential and compartmental extractions on the Cu bioaccessibility in each soil was also measured (Figure S1 to S12) and had a significant effect (Friedman test, $P < 0.05$). The mean Cu bioaccessibility for each soil in sequential extraction were 43.65 ± 0.06 % for Culbin, 42.53 ± 0.02 % for Brechin, 32.27 ± 0.02 % for Insch, and 27.73 ± 0.08 % for St Fergus. Whereas, in the historical, soil, the bioaccessibility was 69.68 ± 0.02 % (Table S1). Comparison of the Cu bioaccessibility in gastric and intestinal phases (D+B) for each soil was tested to find whether there was deference among them (Table S3). A significant difference between the mean Cu bioaccessibility in gastric and intestinal phases were found for historical soil and spiked Culbin soil (p= 0.001 and 0.023, respectively, whereas in other soil no significant differences were observed (p > 0.05). For instance, the average bioaccessibility in gastric and intestinal phases for Culbin soil was 38.05 ± 0.06 and 51.81 ± 0.04 %, while for Brechin soil it was 46.57 ± 0.02 and 46.67 ± 0.02 %, respectively.

3.4. Assessment of Cu bioaccessibility under fasted and fed conditions:

The bioaccessibility of copper was investigated under fasted and fed conditions for Aberlour soil (Figure 2). The bioaccessibility of Cu for sequential and compartmental extractions was also investigated for each soil under fasted and fed conditions (Figure S13 to S16). There was a significant difference within the compartmental extraction between the Cu bioaccessibility under fated and fed conditions (p = 0.002 for saliva, p = 0.003 for gastric, and p = 0.002 for D+B), as in table 6. The mean value for Cu bioaccessibility under fasted condition was 4.79 ± 0.01 % for saliva, 58.8 ± 0.04 % for gastric and 68.79 ± 0.02 % for intestinal phase. Whereas, under fed condition, values were 20.12 ± 0.01 % (saliva), 21.94 ± 0.01 % (gastric) and 92.44 ± 0.02 % (D+B). In the sequential extraction, there was no significant difference between the mean of bioaccessibility among fated and fed conditions (p = 0.825) with mean of 69.68 ± 0.02 % and 70.38 ± 0.01 %, respectively.

Table 6. *P*-values of Paired t-test for the compartmental extraction between the Cu bioaccessibility under fated and fed conditions for Aberlour soil

	Log Sequntial (fasted)	Log Saliva (fasted)	Log Gastric (fasted)	Log D + B (fasted)	All extraction (fasted)
Log Sequ. (fed)	0.825				
Log Saliva (fed)		0.002			
Log Gastric (fed)			0.003		
Log D + B (fed)				0.002	
All extraction (fed)					0.901

3.5. Risk assessment models:

The Risk-Integrated Software for soil Clean-ups (RISC, *Version 4.0*) model was used to estimate human health risk from exposure to contaminated media and to obtain cleanup values (Spence and Walden, 2001). The model considers that a conservative soil ingestion rate for 6 year-old children is 200 mg per day of reasonable maximum exposure and a body weight (BW) of 15 kg. Direct ingestion was the only pathway that was investigated and different values of ingestion soil rate (200, 400, 600, 800 and 1000 mg) were considered to estimate the effect on daily intake or cleanup values. The result showed a correlated relationship between the daily intake values and ingestion soil rate for children ($p<0.001$, $r^2 = 99.9$) and at 200 mg ingestion soil. The model showed daily intake value of 0.04 mg kg^{-1} bw day^{-1}, and cleanup value was 3100 mg kg^{-1}.

3.6. Oral bioavailability and potential Copper intake:

The *in vitro* digestion model used in this experiment showed that it is a fast method and easily to use, can be used as a useful risk assessment tools. The order of magnitude of potential copper intake for children from ingestion of contaminated soil can be calculated based on the Cu bioaccessibility values for sequential extractions, and thus used to approach oral relative bioavailability of Cu in soils (Brandon et al., 2006). The focus on sequential extractions is because it most closely reflects genuine human exposure scenarios.

Even for the historically Cu impacted soils, the bioaccessibility was close to 100 % for a measured dose of 58 mg kg^{-1} in the soil. Therefore, daily intake values (mg kg^{-1} bw day^{-1}) of Cu for child exposed to contaminated soils were calculated taken into account consideration a conservative soil ingestion rate of 200 mg per day of reasonable maximum exposure and a body weight (BW) of 15 kg for 6 year-old children (Spence and Walden 2001, RISC4). The potential daily intake (DI) of Cu in Aberlour (historical contamination) soil (58 mg kg^{-1}) was 0.0008 mg kg^{-1} bw day^{-1} (11.6 µg Cu in 200 mg soil divided by 15 kg the weight of an average child). In spiked Culbin soil, for example, the high Cu bioaccessibility (52.62 %) was found at Cu concentration of 920 mg kg^{-1} showing DI of 0.0123 g kg^{-1} bw day^{-1} for Child. Figure 3 illustrates the daily intake (DI) for each soil (Culbin, Brechin, Insch, St Fergus, and Aberlour) based on Cu concentration at different amount of ingested soil rate.

Therefore, the maximum potential intake of Cu for child exposed to these soils is thus 0.0008 mg kg^{-1}bw day^{-1}. Comparing to TDI value obtained from RIVM- the Netherlands- and RISC4 model (0.14 and 0.04 mg kg^{-1} bw day^{-1}, respectively), it is observed that an average daily intake of Cu per child can be more accurate by using bioaccessibility data of in vitro digestion.

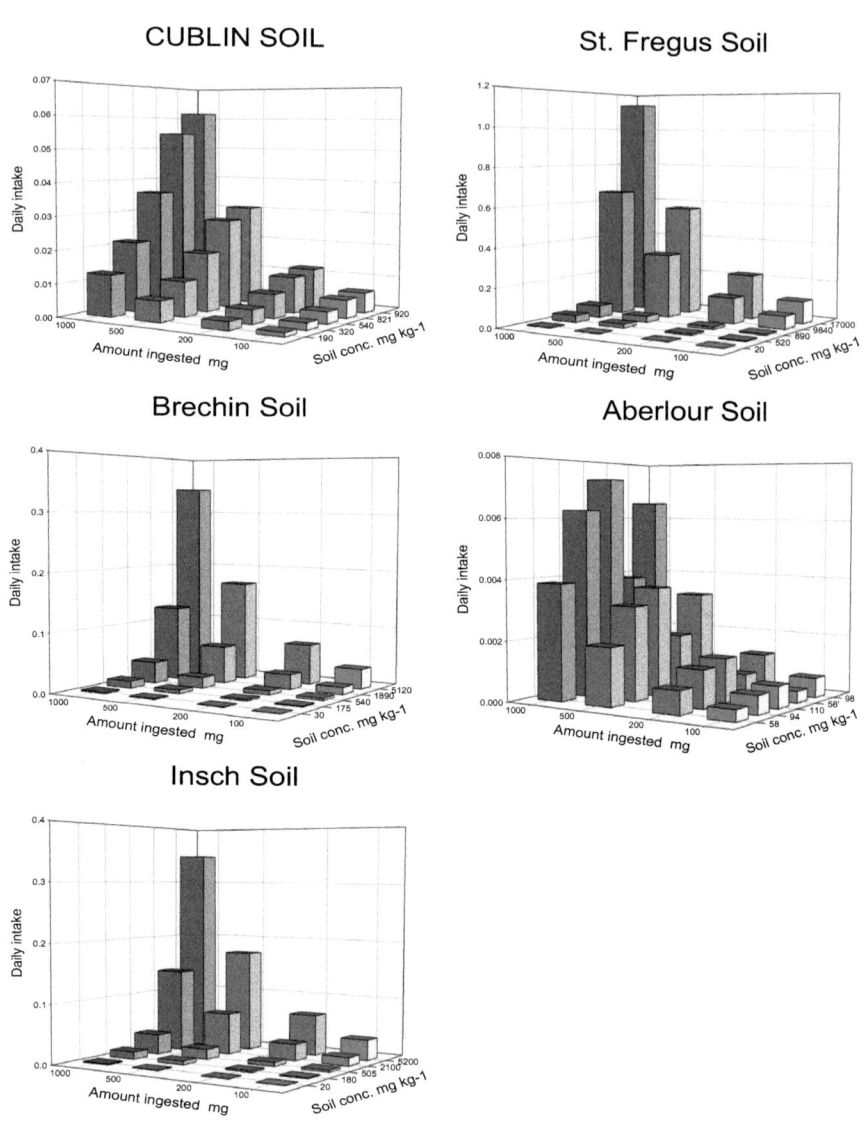

Fig. 3 The daily intake based on Cu conc. for each soil (Culbin, Brechin, Insch, St Fergus, and Aberlour) at different amount of ingested soil rate.

CHAPTER 4

Discussion

Bioaccessibility values depended on the physicochemical attributes of the soils and the nature of the element under investigation. As the results show, there are considerable variations in the Cu bioaccessibility in the gastrointestinal extractions for all soils (Fig. 1). It is acknowledged that copper is mainly retained in soils through ion exchange and specific adsorption mechanisms; additionally, it has a strong affinity for humic compounds (Pouschat and Zagury, 2008). These results suggest that attributes which govern ecological bioavailability may be linked to human exposure also.

In recent studies, the bioaccessibility of different metals (Ni, Cr, Pb, As and Cd) in soil indicated that there was difference in the bioaccessibility (%) between the metals due to differences in origin, sorption behaviour and pH dependence (Ljung *et al.,* 2007). Furthermore, the difference in bioaccessibility of copper is most likely due to the differences in which the copper is bound to the soil constituents, and indeed the nature of these constituents. The effects of soil Cu speciation on the bioaccessibility have been explained by the functionally defined Cu species including water soluble or exchangeable, carbonate bound, the hydroxy complexes (*e.g.*: $CuOH^+$ and $Cu(OH)^{2+}$), and organic matter bound (Liu and Zhao, 2007). It is suggested that when increasing the dose of metal, Cu concentrations in soil pore waters increased with varying degrees in each soil; additionally, the relatively low concentration of Cu is more likely due to the presence of high OM in soil (Liu and Zhao, 2007).

Dawson et al., (2006) reported that partitioning of Cu between solid and solution phases can affect the bioavailability explained by the nature and concentration of the Cu present and also by soil processes that influence mobilisation or transformation). They also highlighted that the main factor for determining the partitioning between solid and solution phases is pH, meaning that low pH increase the total Cu concentrations in soil solution; however, other factors such as oxalate extractable Fe, % sand and % OM might influence the partitioning of Cu.

The pattern with soil seems to be more about texture. Table 2 shows that Culbin Forest has 1.68 % of OM and pH values = 2.00 which commonly retained the least amount of Cu in the solid phase, while St Fergus soil has 8.62 % of OM and pH values = 5.00. This can explain that the Cu bioaccessibility for Culbin soil was higher with increasing Cu concentration in soil than that for St Fergus soil. Furthermore, the presence of high OM in soil could be a major factor in retaining Cu on the soil surface by complexation with phenolic and carboxyl groups (Covelo et al., 2004). Hence, the relatively low concentration of Cu in the digestive solution is probably due to the high OM content in some soil (Dawson et al., 2006).

Although Brechin and Insch soils have pH value > 5.0 and OM % (5.50 and 4.32, respectively), Cu bioaccessibility for both soils were variable even with increasing Cu amendments. The low bioaccessibility can be interpreted by the high pH and OM leading to decrease of Cu insolubility, whereas the high bioaccessibility might due to the degradation of the organic matter with which it associated, e.g. in stomach-like conditions. It also observed that Brechin and Insch soils have silt > 14 % and clay > 11 % comparing to Culbin and St Fergus soils; moreover, the oxalate extractable elements (Al, Fe, Mn and Si) are higher in both Brechin and Insch soils than other soil. This can explain the variation in the Cu bioaccessibility especially with soils that have high Ox. Fe (Table 2) leading to decrease Cu bioaccessibility in the digestive solution with increasing Cu amendments in soil. It is also possible that the bioaccessibility can be influenced by the particle size of soil (180 µg) because the proportion of inorganic carbon (carbonate) increases as particle size of soil decreases leading to an increase in the carbonate minerals concentration in the finer size fractions (Versantvoort et al., 2004).

As an individual compartment, Cu bioaccessibility for all soils were taken out more than the sequential extraction '' whole system '' meaning more of the Cu in soil will be extracted into the body. In recent studies, Cu bioaccessibility from contaminated soils showed a high range in gastric and intestinal phase (25.2 - 94.4 %, and 19.4 - 89.4%, respectively) (Pouschat and Zagury, 2008) indicating that low gastric pH can significantly increase the solubility of metal leading to higher bioaccessibility values and vice versa. (Liu and Zhao, 2007; Intawongse and Dean, 2006). For example, in intestinal phase has pH > 7 causing re-sorption and precipitation of metal after dissolution in the stomach phase (Ruby et al, 1996). Low Cu Bioaccessibility in intestinal phase can be interpreted by the reduced affinity of potential Cu adsorption in intestinal phases as the soil is increasingly amended with Cu (Pouschat and Zagury, 2008). The high bioaccessibility in intestinal phase suggest that Cu is predominantly in a readily exchangeable or soluble form (Balasoiu et al., 2001).

It also was suggested that the importance of the liquid /soil ratio has been overstated indicating by that Cu bioaccessibility in intestinal phase was quite similar (or high in some soils) to that in gastric phase. This variation has been referred to as the H+/soil ratio (mol/g), especially when normalized to stomach acid conditions. Pouschat and

26

Zagury (2008) confirming that the higher Cu bioaccessibility for soils in gastric phase is due to the hydroxide precipitation behaviour and a significant effect pH on the partitioning between solid and solution phases. The Cu(II) in the form of Cu(OH)$^{2+}$ is 100 % in the dissolved form until pH = 5 and then the solubility of Cu(OH)$^{2+}$ rapidly decreased to 12.7% at pH = 5.5; whereas, at pH > 6.0 all Cu(OH)2 in the system transformed to CuO solid with 99 % of Cu(II) (Liu and Zhao, 2007). Thus, it is observed that the compartmental extractions are always more harsh, and showed complexes for understanding the higher bioaccessibilities in gastric and intestinal parts that the whole system.

Comparing spiked soil with historical Aberlour soil, Aberlour soil released significantly less than the spiked soils. Cu concentration in Aberlour soil ranged from 58 to 110 mg kg^{-1} while spiked soils had high amendments of Cu which can cause an overestimation of the Cu concentration solution when compared to historically contaminated soils (Zhang et al., 2004). The higher Cu bioaccessibility (106 %) in Aberlour soil was found at the lower Cu concentration 58 mg kg^{-1} and the lower was at high concentration. This can be explained by the factors mentioned above. The presence of food components with Aberlour soil shows significant effects on the bioavailability process of ingested contaminants from soil in the compartmental extractions. An explanation as to the differences in Cu bioaccessibility value under fasted and fed conditions has been attempted. Additionally, it may be strongly dependent on the food source (Versantvoort et al., 2005) and this might attribute to the fact that in vitro method of RIVM, which used infant formula selected in this study, did not have standardised fed model comparing with other models (Van de Wiele et al. 2007). In addition, the presence of nutrition in the gastrointestinal juice means more presence of dissolved organic matter, and thus providing more complication niches for contaminant in organic material solution rather than in a freely absorbable condition (Van de Wiele et al. 2007).

The daily intake can be calculated based on the bioaccessibility data in which children can expose to the potential risk from contaminated soil. It was observed that very few *in vivo* studies on soil samples for Copper have been reported in the literature, and the reliable estimates of relative bioavailability have not been obtained (Kelley et al. 2002). Up to date, no in vitro gastrointestinal protocols have been successfully validated with in vivo data for Cu (Pouschat and Zagury, 2008). Furthermore, the use of 100% bioaccessibility default value (in risk assessments) appears to be overly as these results demonstrate. The RISC4 assessment model proposed cleanup value 3100 mg kg^{-1} for children (15 kg) with ingestion soil rate (200 mg) and ingestion daily of soil (0.04 mg kg^{-1} day^{-1}). Using bioaccessibility data of in vitro digestion model, the high Cu bioaccessibility (100 %) which might cause risk to children was found on Aberlour soil (58 mg) giving an average daily intake of Cu per child base on (0.0008 mg kg^{-1}bw day^{-1}). Moreover, different spiked soil showed different values of daily intake for copper (Fig. 3). Therefore, in vitro data can be more accurate than risk assessment models for copper, and thus the hypothesis that risk assessment models for copper in different soils are more conservative than empirical data is rejected.

CHAPTER 5

Conclusion

The *in vitro* digestion model (RIVM) used in this experiment is a fast and relatively inexpensive method to simulate the release of Cu from contaminated soil and for assessing potential risk of contaminants to humans. The bioaccessibilities of copper in spiked soil samples under fasted condition were variable indicating that the bioaccessibility correlated with the soil texture while soil pH has a lesser effect. Additionally, the bioaccessibility in historically contaminated soils was proportionally similar to that for spiked soils. Assays using fasted scenarios showed different results from fed assays. However, it is recommended that in vitro models need additional studies and development for providing a standardized method for fed conditions to be used in human health risk assessment. Measurements of metal assimilation using *in vitro* digestion model through direct ingestion represent an important pathway for exposure to contaminated soils. Furthermore, empirical data of copper for different soils appears to be more conservative than that from risk assessment models.

ACKNOWLEDGEMENTS

I gratefully thank Dr. Graeme Paton for his supervision, support and his efforts throughout all steps of this project. Thanks are extended to Al Hadrami, Hani for his guidance and assistance. I would like to acknowledge and thank everyone who advised and assisted me during my work. My profound gratitude goes to God for his Grace, then to my parents and family for their support.

Khaled Ali Alsaleh

2015, Riyadh, kingdom of Saudi Arabia

REFERENCES

Baars, A.J., Theelen, R.M.C., Janssem, P.J.C.M., Hesse, J.M., van Apeldoorm, M.E., Meijerink, M.C.M., Verdam, L., Zeilmaker, M.J. (2001) Re-evaluation of human-toxicological maximum permissible risk levels. RIVM report no. 711701025, National Institute of Public Health and the Environment, p 62-65. Available at http://www.rivm.nl/en/

Balasoiu, C. F., Zagury, G. J., and Deschênes, L. (2001). "Partitioning and speciation of chromium, copper, and arsenic in CCAcontaminated soils: Influence of soil composition." Sci. Total Environ., 280, 239–255.

Brandon, E. F.A., Oomen, A. G., Rompelberg, C. J.M., Versantvoort, C. H.M., Engelen, J. G.M. van, Sips A. J.A.M. (2006) Consumer product in vitro digestion model: Bioaccessibility of contaminants and its application in risk assessment. Regulatory Toxicology and Pharmacology 44, 161–171

Bremner, I, Beattie, JH. (1995). Copper and zinc metabolism in health and disease: Speciation and interactions. Proc Nutr Soc 54:489-499.

Calabrese, E. J., Stanek, E. J., James, R. C., Roberts, S. M. (1997) Soil ingestion: a concern for acute toxicity in children. Environ Health Perspect;105: 1354– 8.

Casteel, S. W., Cowart, R. P., Weis, C. P., Henningsen, G. M., Hoffman, E., et al. (1997) Bioavailability of Lead to Juvenile Swine Dosed with Soil from the Smuggler Mountain NPL Site of Aspen, Colorado. Fundam. Appl. Toxicol., 36, 177-187.

Covelo, E.F., Andrade, M.L., Vega, F.A., (2004) Heavy metal adsorption by humic umbrisols: selectivity sequences and competitive sorption kinetics. Journal of Colloid and Interface Science 280, 1-8.

Davis, A., Bloom, N. S., and Que Hee, S. S. (1997) The environmental geochemistry and bioaccessibility of mercury in soils and sediments: A review. Risk Anal. 17:557–569.

Dawson, J.J.C., Campbell, C.D., Towers,W., Cameron, C.M., Paton G.I. (2006) Linking biosensor responses to Cd, Cu and Zn partitioning in soils. Environmental Pollution; 142, 493-500

Ellickson, K.M., Meeker, R.J., Gallo, M.A., Buckley, B.T. and Lioy, P.J. (2001) Oral bioavailability of lead and arsenic from a NIST standard reference soil material, Environ. Contam. Toxicol. (40), 128–135.

Environment Agency (2005) Report on the International Workshop on the Potential Use of Bioaccessibility Testing in Risk Assessment of Land Contamination. Science Report SC040054. In http://www.environmentagency.gov.uk/subjects/landquality/113813/1283985/?version=1&lang=_eIntawongse, M., Dean, J. R. (2006) In-vitro testing for assessing oral bioaccessibility of trace metals in soil and food samples. Trends in Analytical Chemistry, Vol. 25, No. 9.

Intawongse, M., Dean, J. R. (2006) In-vitro testing for assessing oral bioaccessibility of trace metals in soil and food samples. Trends in Analytical Chemistry, Vol. 25, No. 9.

Intawongse, M., Dean, J. R. (2008) Use of the physiologically-based extraction test to assess the oral bioaccessibility of metals in vegetable plants grown in contaminated soil. Environmental Pollution (152) 60-72.

Juhasz, A. L., Smith, E., Weber, J., Rees, M., Rofe, A., Kuchel, T., Sansom, L., Naidu, R. (2007) In vitro assessment of arsenic bioaccessibility in contaminated (anthropogenic and geogenic) soils Chemosphere (69) 69–78.

Kelley, M. E., Brauning, S. E., Schoof, R. A., and Ruby, M. V. (2002) Assessing oral bioavailability of metals in soil, Batelle Press, Columbus, Ohio. Based on reference by Intawongse, M., Dean, J. R. (2008).

Liu, R., Zhao, D. (2007) In situ immobilization of Cu(II) in soils using a new class of iron phosphate nanoparticles. Chemosphere (68)1867–1876.

Ljung, K., Oomen, A., Duits, M., Selinus, O. and Berglund, M. (2007) Bioaccessibility of metals in urban playground soils. Environmental Science and Health Part A 42, 1241–1250.

Oomen, A.G., Hack, A., Minekus, M., Zeijdner, E., Cornelis, C., Schoeters, G., Verstraete, W., Van de Wiele, T., Wragg, J., Rompelberg, C.J.M., Sips, A.J.A.M. and Van Wijnen, J.H. (2002) Comparison of five in vitro digestion models to study the bioaccessibility of soil contaminants. Environ. Sci. Technol. 36, 3326–3334.

Oomen, A.G., Rompelberg, C.J.M., Bruil, M.A., Dobbe, C.J.G., Pereboon, D.P.K.H., Sips, A.J.A.M. (2003) Development of an in vitro digestion model for estimating the bioaccessibility of soil contaminants. Arch. Environ. Contam. Toxicol. 44, 281–287.

Paustenbach, DJ (2000) The Practice Of Exposure Assessment: A STATE-OF- THE-ART REVIEW. Toxicol Environ Health. Part B 3:179–291

Pouschat, P.a and Zagury, G. J. (2008) Bioaccessibility of Chromium and Copper in Soils near CCA-Treated Wood Poles. Practice Periodical of Hazardous, Toxic, and Radioactive Waste Management ©ASCE, Vol. 12, No. 3-216–223.

Rodriguez, R.R., Basta, N.T., Casteel, S.W., Pace, L.W. (1999) An in-vitro gastro-intestinal method to assess bioavailable arsenic in contaminated soils and solid media. Environ. Sci. Technol. 33, 642–649.

Rotard, W., Christmann, W., Knoth, W. and Mailahn, W. (1995) Bestimmung der resorptionsverfu"gbaren PCDD/PCDF aus Kieselrot. UWSF-Z Umweltchem Okotox 7:3–9. As referenced in Oomen's paper 2003.

Ruby, M.V., Davis, A., Schoof, R., Eberle, S. and Sellstone, C.M. (1996) Estimation of lead and arsenic bioavailability using a physiologically based extraction test. Environ. Sci. Technol. 30, 422–430.

Ruby, M.V., Schoof, R., Brattin, W., Goldade, M., Post, G., Harnois, M., Mosby, D.E., Casteel, S.W., Berti, W., Carpenter, M., Edwards, D., Cragin, D. and Chappell, W. (1999) Advances in evaluating the oral bioavailability of inorganics in soil for use in human health risk assessment. Environ. Sci. Technol. 33, 3697–3705.

Spence, L. R., Walden, T. (2001) BP's Risk-Integrated Software for Cleanups (RISC); RISC User's Manual Version 4.0

Van de Wiele, T. R., Oomen, A. G., Wragg, J., Cave, M., Minekus, M., Hack, A., Cornelis, C., Rompelberg, C. J. M., De Zwart, L. L., Klinck, B., Wijnen, J. V., Verstraete, W. And Sips, A. J.A.M. (2007) Comparison of five in vitro digestion models to in vivo experimental results: Lead bioaccessibility in the human gastrointestinal tract. Journal of Environmental Science and Health Part A 42, 1203–1211

Versantvoort, C. H. M., Oomen, A. G., Van de Kamp, E., Rompelberg, C. J. M., Sips, A. J. A. M. (2005) Applicability of an in vitro digestion model in assessing the bioaccessibility of mycotoxins from food. Food and Chemical Toxicology (43) 31–40.

Versantvoort, C.H.M., Van de Kamp, E., Rompelberg, C.J.M., (2004) Development of an in vitro digestion model to determine the bioaccessibility of contaminants from food. Report no. 320102002, Available from <http:/www.rivm.nl/en/>, National Institute for Public Health and the Environment, Bilthoven, The Netherlands.

Wittsiepe, J., Schrey, P., Hack, A., Selenka, F. and Wilhelm, M. (2001) Comparison of different digestive tract models for estimating bioaccessibility of polychlorinated dibenzo-p-dioxins and dibenzofurans (PCDD/F) from red slag 'Kieselrot'. Int. J. Hyg. Environ. Health 203, 263–273.

World Health Organization, Geneva, Switzerland. (WHO, 1996) Guidelines for drinking-water quality, 2nd ed. Volume 2, Health criteria and other supporting information. Available on the website: http://www.popline.org/docs/1460/128859.html

Zhang, H., Lombi, E., Smolders, E., McGrath, S., (2004) Kinetics of Zn release in soils and prediction of Zn concentration in plants using diffusive gradients in thin films. Environmental Science and Technology 38 (13), 3608-3613.

Supplementary data

Table S1: The bioaccessibility % of Cu for all soils + SE

Type of soil	Total conc. (mg kg-1)	Bioaccessibility %			
		Sequential	Saliva	Gastric	D + B
Cublin soil	190	32.10 ± 0.02	3.73 ± 0.04	24.51 ± 0.06	33.73 ± 0.01
	320	40.25 ± 0.02	3.12 ± 0.02	30.91 ± 0.11	44.96 ± 0.09
	540	41.26 ± 0.01	6.11 ± 0.09	39.81 ± 0.03	51.62 ± 0.04
	821	52.01 ± 0.04	5.98 ± 0.03	50.98 ± 0.06	69.74 ± 0.02
	920	52.62 ± 0.21	6.54 ± 0.06	44.05 ± 0.03	59.00 ± 0.03
	mean	**43.65± 0.06**	**5.10 ± 0.05**	**38.05 ± 0.06**	**51.81 ± 0.04**
Brechin soil	30	41.33 ± 0.01	3.15 ± 0.01	33.98 ± 0.01	42.00 ± 0.01
	175	39.55 ± 0.01	2.12 ± 0.01	35.78 ± 0.02	39.33 ± 0.01
	540	48.27 ± 0.02	2.24 ± 0.01	50.45 ± 0.03	51.92 ± 0.01
	1890	45.26 ± 0.02	2.52 ± 0.01	57.11 ± 0.16	55.24 ± 0.01
	5120	38.26 ± 0.03	2.09 ± 0.05	55.56 ± 0.68	44.89 ± 0.06
	mean	**42.53 ± 0.02**	**2.42 ± 0.02**	**46.58 ± 0.18**	**46.68 ± 0.02**
Insch Soil	20	47.66 ± 0.01	4.28 ± 0.01	31.50 ± 0.01	32.10 ± 0.01
	180	26.13 ± 0.01	1.94 ± 0.01	36.38 ± 0.02	35.10 ± 0.01
	505	32.34 ± 0.01	1.92 ± 0.01	45.67 ± 0.02	44.28 ± 0.04
	2100	29.50 ± 0.02	1.43 ± 0.02	55.43 ± 0.22	34.91 ± 0.10
	5200	25.73 ± 0.03	1.18 ± 0.06	48.36 ± 0.25	28.90 ± 0.70
	mean	**32.27 ± 0.02**	**2.15 ± 0.02**	**43.47 ± 0.10**	**35.06 ± 0.17**
St. Fregus Soil	20	51.15 ± 0.01	4.73 ± 0.01	29.93 ± 0.01	63.00 ± 0.01
	520	10.75 ± 0.01	1.15 ± 0.01	52.60 ± 0.06	49.53 ± 0.03
	890	32.00 ± 0.12	2.38 ± 0.03	163.17 ± 0.01	96.67 ± 0.10
	9840	23.65 ± 0.15	0.91 ± 0.06	65.07 ± 1.54	31.46 ± 1.02
	17000	21.10 ± 0.10	1.00 ± 0.27	67.13 ± 2.47	28.44 ± 0.51
	mean	**27.73 ± 0.08**	**2.03 ± 0.08**	**75.58 ± 0.82**	**53.82 ± 0.33**
Aberlour Soil	58	105.69 ± 0.03	8.61 ± 0.00	87.17 ± 0.04	108.10 ± 0.00
	94	62.33 ± 0.02	4.69 ± 0.02	54.38 ± 0.07	55.21 ± 0.02
	110	36.28 ± 0.01	2.07 ± 0.01	30.93 ± 0.03	40.42 ± 0.03
	58	69.75 ± 0.01	3.52 ± 0.01	57.84 ± 0.02	64.86 ± 0.02
	98	74.34 ± 0.01	5.05 ± 0.01	63.67 ± 0.02	75.37 ± 0.02
	mean	**69.68± 0.02**	**4.79 ± 0.01**	**58.80 ± 0.04**	**68.79 ± 0.02**
Aberlour Soil Under Fed Condition	58	106.50 ± 0.01	30.78 ± 0.01	39.06 ± 0.01	138.31 ± 0.02
	94	67.03 ± 0.01	19.52 ± 0.01	23.77 ± 0.01	86.81 ± 0.02
	110	45.30 ± 0.01	12.91 ± 0.01	12.62 ± 0.01	56.56 ± 0.03
	58	61.87 ± 0.01	20.35 ± 0.01	15.25 ± 0.01	79.66 ± 0.01
	98	71.17 ± 0.01	17.04 ± 0.01	18.99 ± 0.01	100.84 ± 0.01
	mean	**70.37 ± 0.01**	**20.12 ± 0.01**	**21.94± 0.01**	**92.44 ± 0.02**

Table S2. P-values of Paired t-test for the extraction versus the Cu conc. For all soils.

Copper	Conc. 0	Conc. 50	Conc. 250	Conc. 1000	Conc. 2500
Sequntial	0.089	0.907	0.018	0.027	0.027
Saliva	0.114	0.033	0.004	0.009	0.011
Gastric	0.138	0.563	0.057	0.016	0.016
D + B	0.127	0.533	0.028	0.034	0.028

Table S3. P-values of t-test for the Cu bioaccessibility in gastric and intestinal phases (D+B) for each soil

		Cu bioaccessibility in Gastric				
		Culbin soil	Brechin soil	Insch soil	Fergus soil	Aberlour soil
	Culbin soil	0.001				
	Brechin soil		0.814			
D + B	Insch soil			0.153		
	Fergus soil				0.383	
	Aberlour soil					0.023

Table S4. P-values of Paired t-test for Cu bioaccessibility of the compartmental extraction versus the sequential extraction for all soils.

	Log Saliva	Log Gastric	Log D + B
Log Sequntia	0.000	0.049	0.010
Log Saliva		0.000	0.000
Log Gastric			0.664

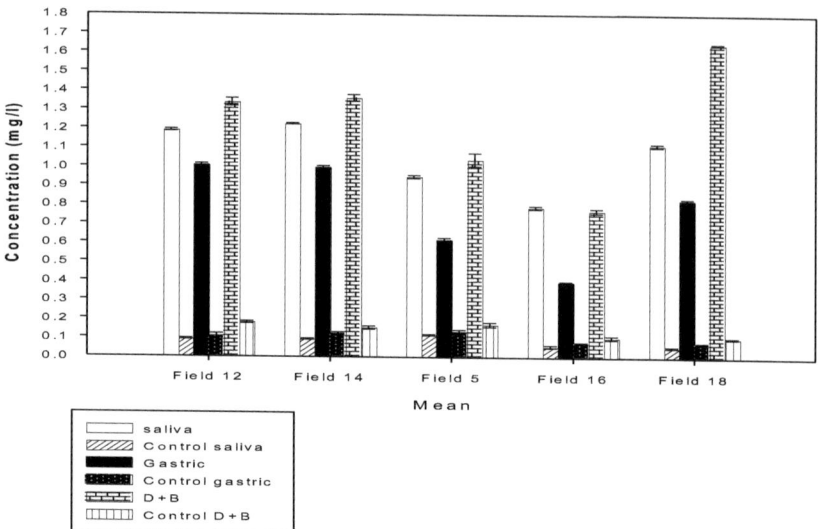

Fig. S1. The mean of replicated and controlled samples of each Aberlour soil for compartmental extraction under fed condition.

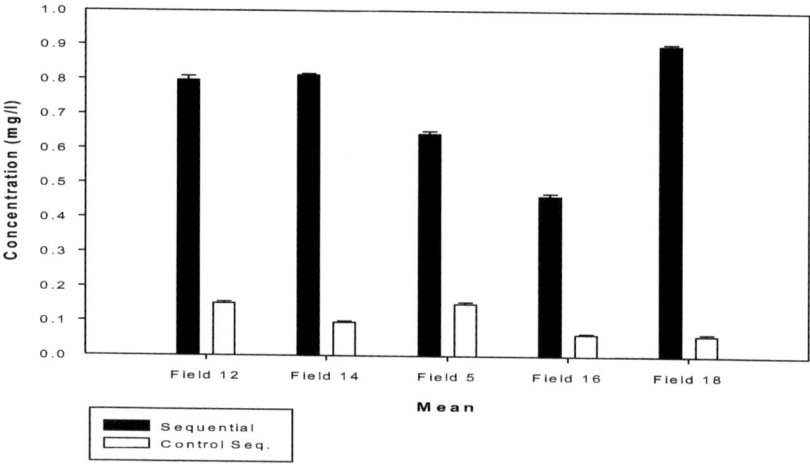

Fig. S2. The mean of replicated and controlled samples of each Aberlour soil for sequential extraction under fed condition.

Under Fasted Condition:

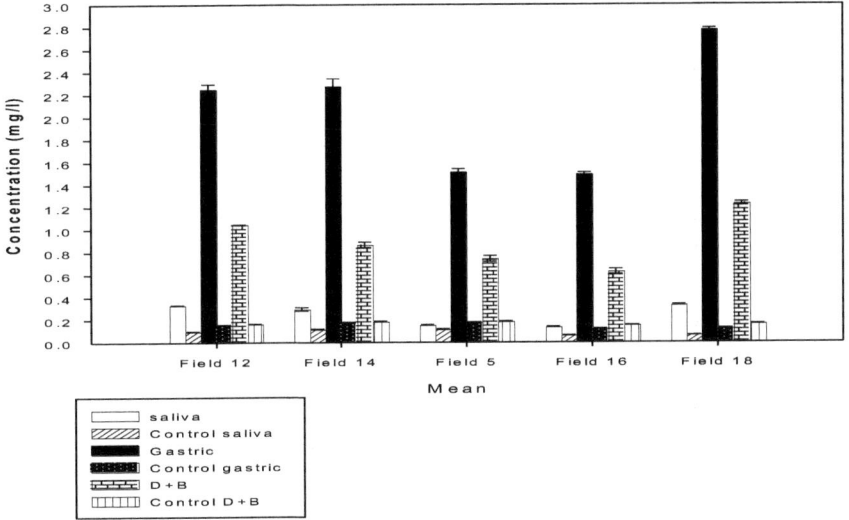

Fig. S3. The mean of replicated and controlled samples of each Aberlour soil for compartmental extraction under fasted condition.

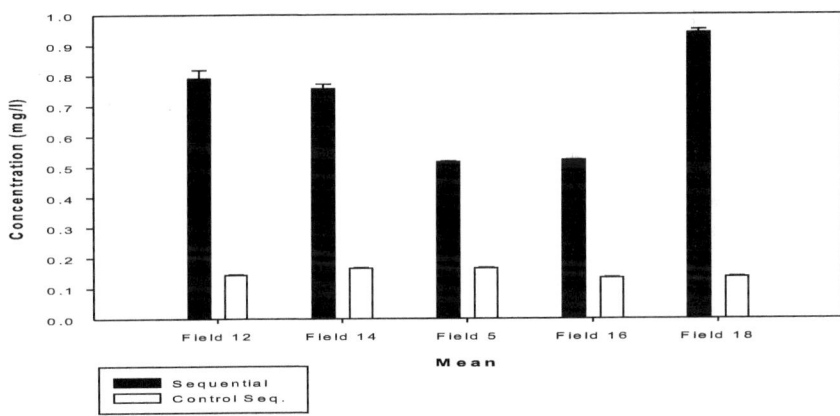

Fig. S4. The mean of replicated and controlled samples of each Aberlour soil for sequential extraction under fasted condition.

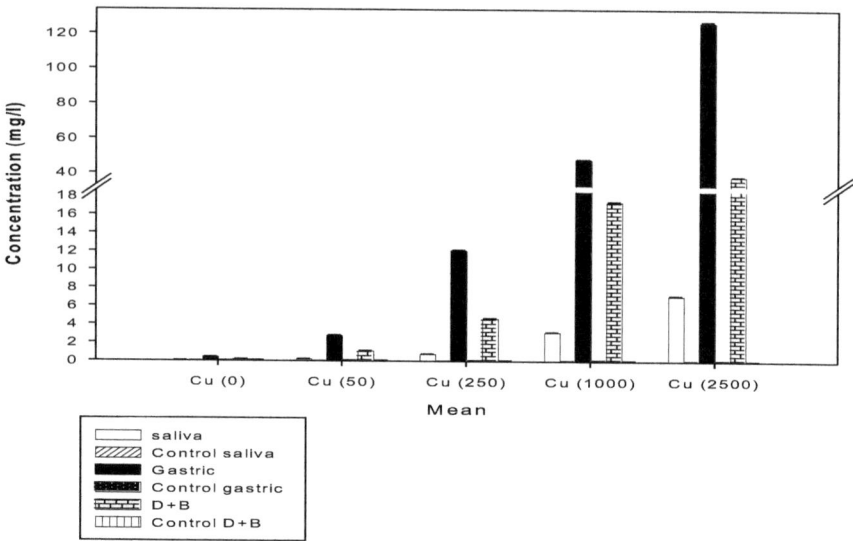

Fig. S5. The mean of replicated and controlled samples of each Brechin soil for compartmental extraction under fasted condition.

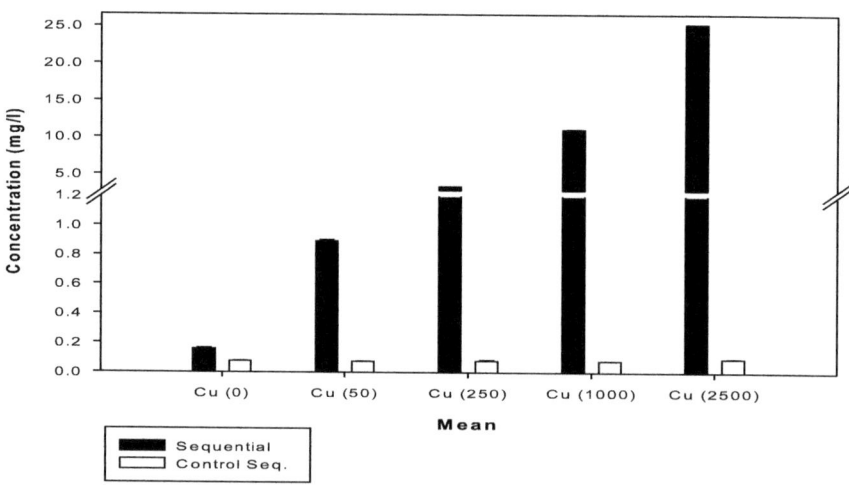

Fig. S6. The mean of replicated and controlled samples of each Brechin soil for sequential extraction under fasted condition.

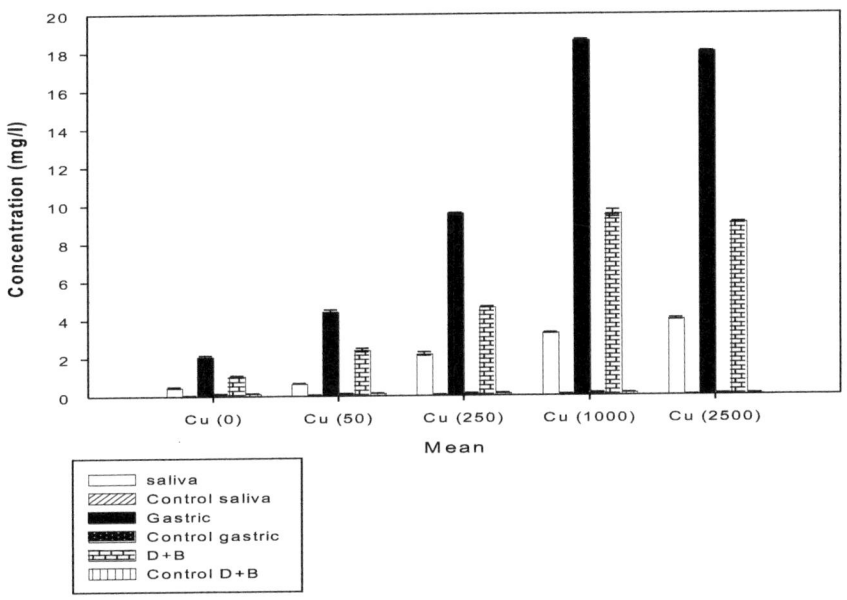

Fig. S7. The mean of replicated and controlled samples of each Cubin soil for compartmental extraction under fasted condition.

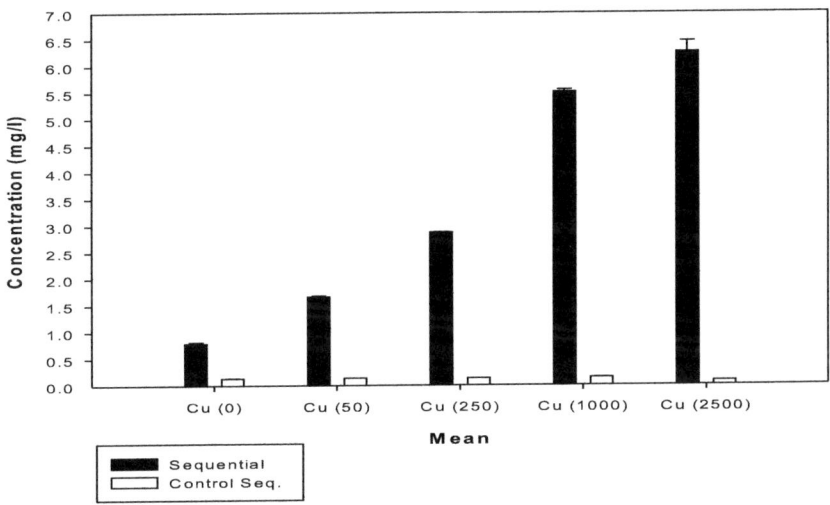

Fig. S8. The mean of replicated and controlled samples of each Cubin soil for sequential extraction under fasted condition.

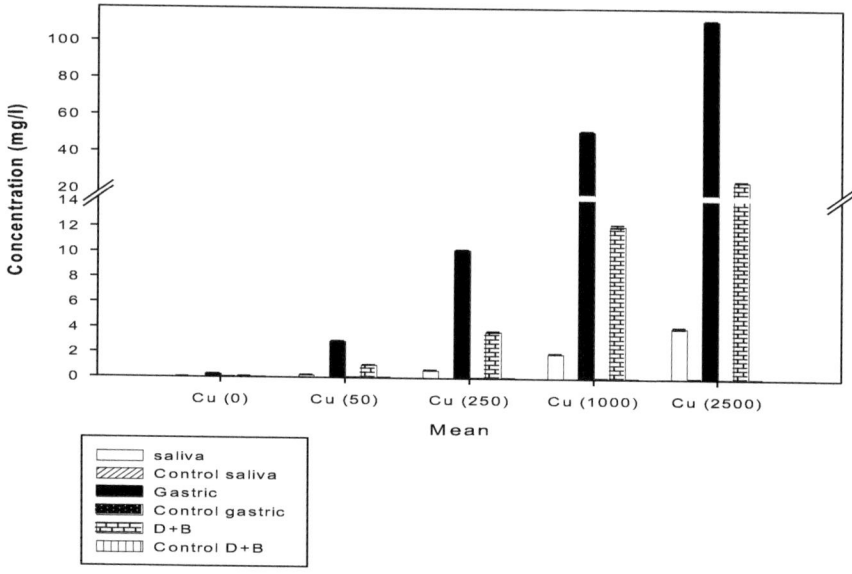

Fig. S9. The mean of replicated and controlled samples of each Insch soil for compartmental extraction under fasted condition.

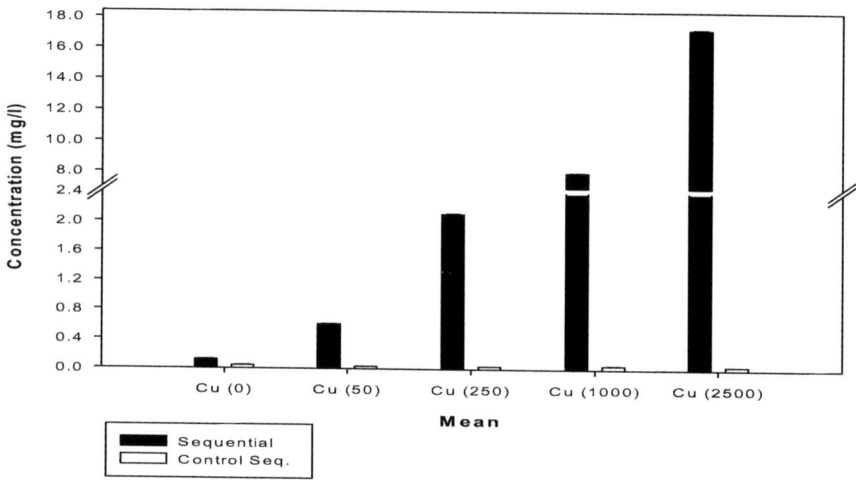

Fig. S10. The mean of replicated and controlled samples of each Insch soil for sequential extraction under fasted condition.

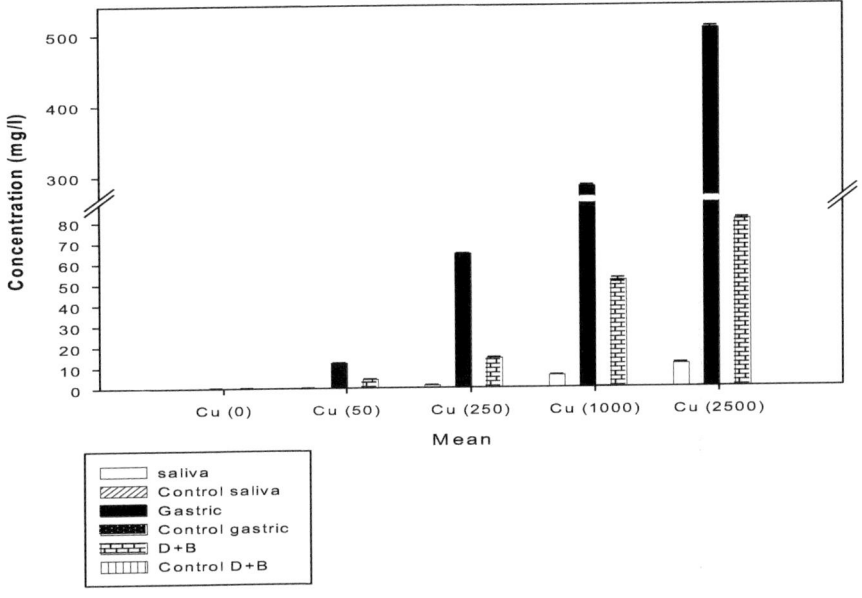

Fig. S11. The mean of replicated and controlled samples of each St. Fregus soil for compartmental extraction under fasted condition.

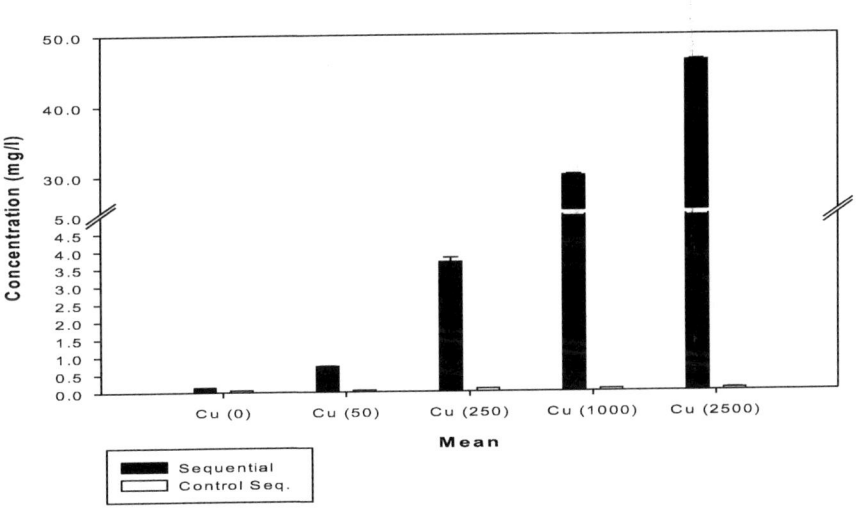

Fig. S12. The mean of replicated and controlled samples of each St. Fregus soil for sequential extraction under fasted condition.

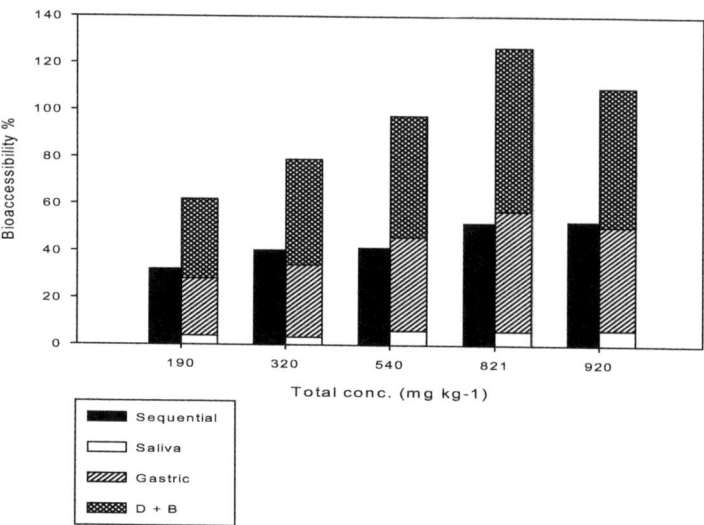

Fig. S13. Sequential and compartmental extractions for Cublin soil spiked with Copper under fasted condition.

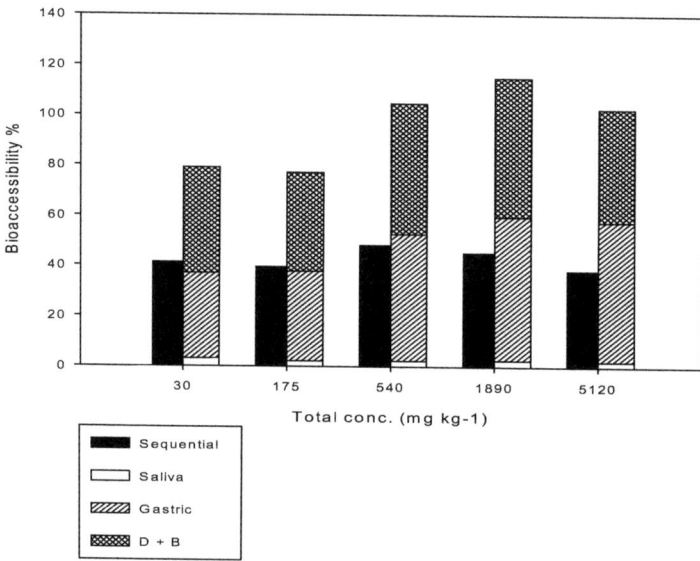

Fig. S14. Sequential and compartmental extractions for Brechin soil spiked with Copper under fasted condition.

42

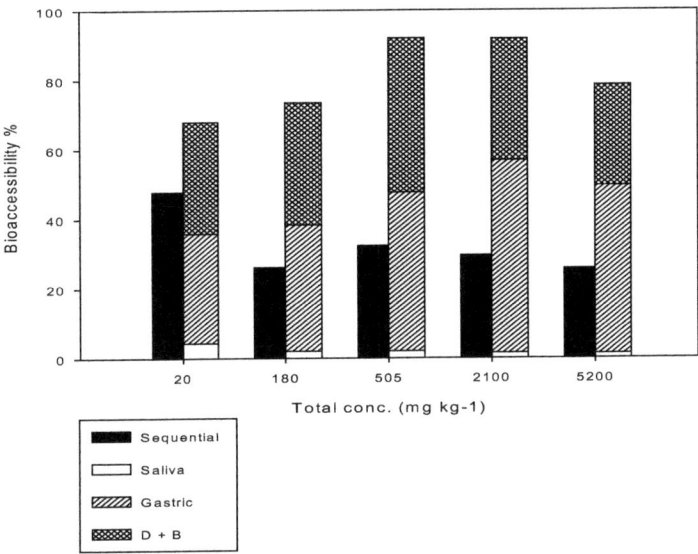

Fig. S15. Sequential and compartmental extractions for Insch soil spiked with Copper under fasted condition.

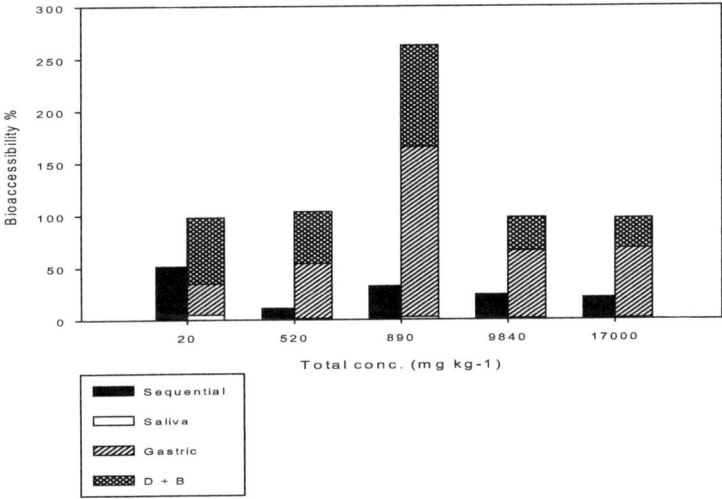

Fig. S16. Sequential and compartmental extractions for St. Fregus soil spiked with Copper under fasted condition.

43